Selected Works
Of
Zarrar Nashman

Introduction

The following are a few selected works of myself which express some of my contemporary theories and ideas. This book is not meant to be used as a rigorous text but rather a mixture of many competing ideas that we may see in modern science. One of my works explored in this book is theoretical and the reader is encouraged to add to the ideas expressed with his or her knowledge and that is the first chapter. Two of the works cited in this book are experiments which have not been in the ideal environments but have been manipulated to mimic one, which is more evident throughout the text. The book's final chapter is a personal piece which I added from the time I had to write an essay about myself and why I deserve to be in my college's honors program.

Selected Works

Contents:

Pg.

3) Electromagnetic Induction for harnessing the earth's constantly shifting magnetic properties in order to access infinite energy.

14) Investigating the relationship between the sinusoidal patterns of a pendulum's position, velocity, acceleration, jerk, and jounce.

24) Investigating the effect of magnesium on photosynthesis

30) Honors program acceptance Essay

Electromagnetic Induction for harnessing the earth's constantly shifting magnetic properties in order to access infinite energy.

Abstract:

It is well known to us the different methods of energy production that we have today. Despite those alternatives, our energy production methods are still very inefficient and primitive for the most part and not helpful to the environment. This paper will look into one different alternatives that we can turn to, which is magnetism. The concept behind harnessing energy through magnetism is relatively straight forward and since it has no potential bi products, it is non pollutant. The concept behind how using this method will be addressed in detail as we proceed further into the paper. Starting from the things we use every day, like our electric toothbrush to the vast electric generators, we use magnetic induction. In a nutshell, when a magnet passes through an induction coil, typically made of copper, a certain amperage of current is produced which can then be used for our daily energy needs. Applying this concept to the earth will be a little tricky, however, this is arguably the most efficient and eco-friendly system of producing renewable energy, even in theoretical terms. We will be looking into the real world application of this idea and the prospects of actually being able to make this work. The whole idea revolves around the source, which in this case, is the earth's core. Controversies do and will continue to exist against this idea coming into play, however, it is important to note that even successful methods of energy production has controversies.

Let us briefly take a look at what magnets are. In most objects, all the atoms are in balance with half of the negatively charged particles called electrons, spinning in one direction and half spinning in the other direction. Magnets are quite different with the atoms at one end having electrons spinning in one direction while those at the other end spin in the opposite direction. We call one end of the magnet the North (N) pole and the other end the South (S) pole. The force of the magnetic field flows from the North pole to the South pole. It is theoretically possible to make magnets by hand by constantly stroking a metal in the same direction for a long period of time[1]. Another, more convenient way to make a magnet is through electricity. Each electron in a metal is surrounded by a force called an electric field. When an electron moves, it creates a secondary field called a magnetic field. When electrons are made to flow in _____ a current through a conductor, such as a piece of metal or a coil of wire, the conductor becomes a temporary magnet, also called an electromagnet since it is being created by electricity. Just as we can make magnets from electricity we can also use magnets to make electricity. A magnetic field pulls and pushes electrons in some objects near them to make them move. Metals, like copper, have electrons that are moved easily and can be readily moved from their orbits. If a magnet is moved quickly through a coil of copper wire, electrons move and electricity is made. This process is called electromagnetic induction. When an electrical wire cuts across magnetic lines of force, a current is produced in the wire.

This also corresponds with Faraday's law, which states that a change in the magnetic field is positively correlated with the amount of voltage or electromotive force produced. We know this because the current is detected by navigating the needle on an ammeter or multimeter, which are instruments that can measure electric current in wires. The same result is obtained when a magnet is moved in and out of coils of wire. It does not matter if the magnet is moved or if the coils of wire are moved. The important thing is that there is motion within the magnetic field which would change the magnetic lines of force passing through the coil. A graphical correlation of Faraday's Law would look something like this:

---

1 "Making a Magnet." University of Leicester, n.d. Web. 01 Nov. 2016. <https://www.le.ac.uk/se/centres/sci/selfstudy/mam9.htm>.

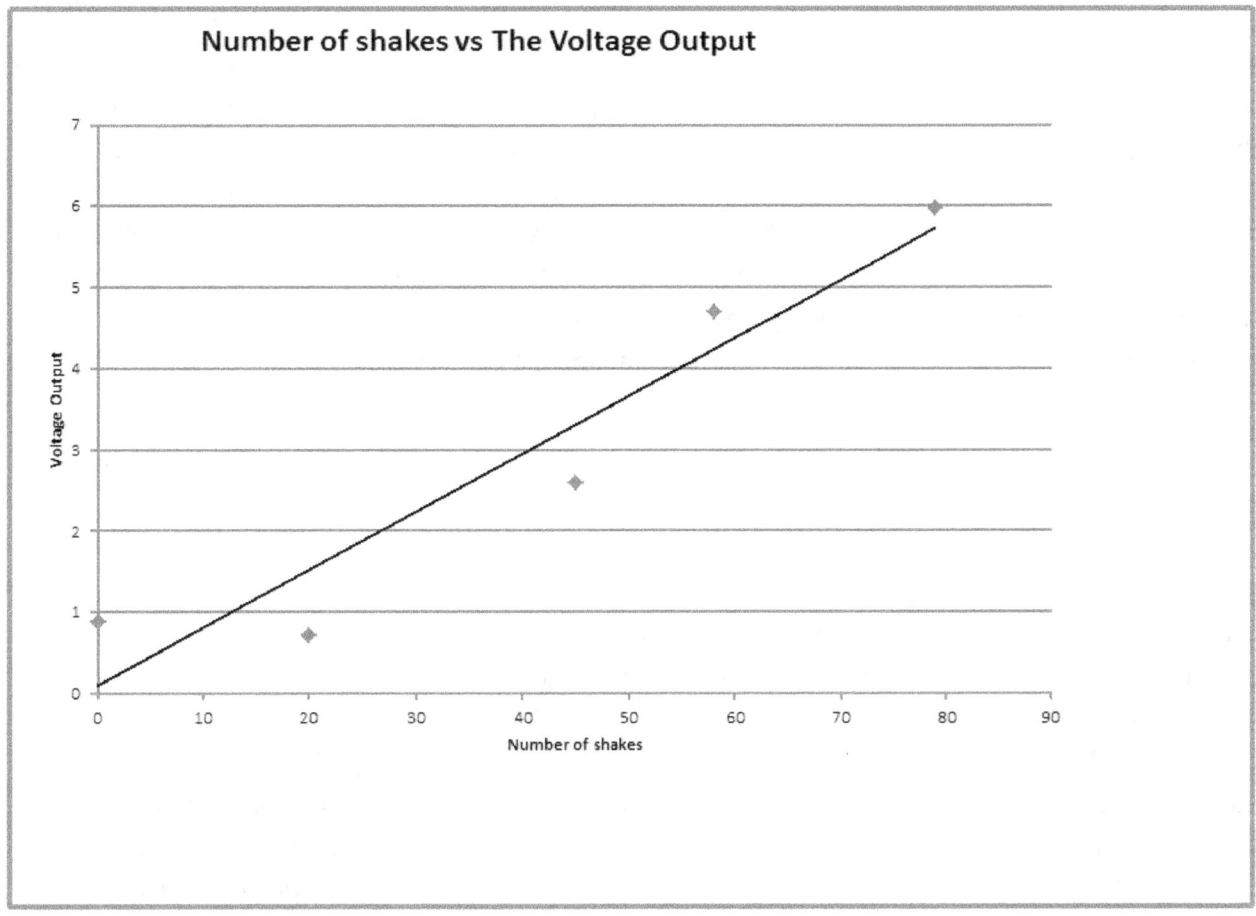

Sites.suffolk.edu   . Suffolk University, 29 Oct. 2013. Web.

The graph above was an experiment at Suffolk University involving a magnet placed within a compact plastic tube which was sealed on both sides on both ends with enough room for the magnet to move up and down the tube with an led attached, essentially making a torch. Then a coil was wrapped around it and a voltmeter was connected to the circuit in order to record the voltage or emf produced . As we can see, the scatter plot above shows a clear positive correlation between the voltage and the number of shakes. This concludes that motion interrupting the field does produce a voltage. And concluding from this experiment, we can also say that the more frequently the quicker the field is disrupted, the more voltage would be produced. This partially proves Faraday's law as well since electricity is being produced.

Let us have a look at how this concept can be applied to our earth. The earth's magnetic field, also known as the geomagnetic field, is the magnetic field that extends from the Earth's interior out into space, where it meets the solar wind, a stream of charged particles emanating from the Sun. Our planet's magnetic field is believed to be generated deep down in the Earth's core. Nobody has ever taken the mythical journey to the centre of the Earth, but by studying the way shockwaves from earthquakes travel through the planet, physicists have been able to carefully work out its likely structure. Right at the heart of the Earth is a solid inner core, two thirds of the size of the Moon and composed primarily of iron. At a hellish 5,700°C, this iron is just as hot as the Sun's surface, but the crushing pressure caused by gravity prevents it from becoming liquid.

Surrounding this is the outer core, a 2,000 km thick layer of iron, nickel, and small quantities of other metals. Lower pressure than the inner core means the metal here is fluid. Differences in temperature, pressure and composition within the outer core cause convection currents to occur in the molten metal as the cool, dense matter sinks whilst warm, less dense matter rises. The Coriolis force, resulting from the Earth's spin, also causes swirling whirlpools in the outer core. This flow of liquid iron generates electric currents, which in turn produce magnetic fields. Charged metals passing through these fields go on to create electric currents of their own, and since liquid molecules are constantly changing their positions, the cycle continues. This self-sustaining loop is known as the geodynamo. It is this geodynamo that will allow the movement of the magnetic body, which is key in this process. The actual process of keeping the fluid body liquid may be more complicated than that and perhaps the motion of the fluid has more than primarily the Coriolis force. This contributors for the motion will not affect our results as all we need is the motion itself.

The spiralling caused by the Coriolis force means that separate magnetic fields created are roughly aligned in the same direction, their combined effect sums up to produce one vast magnetic field engulfing the planet. Lets note that convection currents, as described earlier, occur in the outer core, which is fluid by nature. As we have addressed earlier that in order for induction to happen, there needs to be some disruption in the magnetic flux-area covered by a magnetic field-or when the number of magnetic field lines that pass through the loop is changing. And in order for that to

happen, either the coil or the magnet has to move. Since liquids are constantly moving, we can safely assume that mobility will not act as a problem, as long as we have a coil surrounding it.

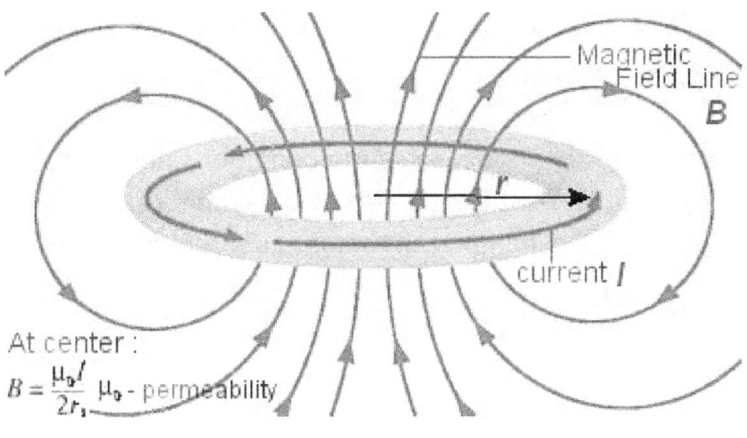

2

Liquid magnets, more commonly known as ferrofluids, have the fluid properties of a liquid and the magnetic properties of a solid. The ones that are created artificially contain tiny magnetic particles suspended in a solution. A ferrofluid consists of nano particles called magnetite that give them their magnetic properties. A ferrofluid has an application where the can behave as a liquid O-ring where a rotating shaft enters either a low- or high-pressure chamber. The ferrofluid is held in place by permanent magnets and forms a very tight seal, eliminating most of the friction produced in a traditional mechanical seal that use solid instead. These rotating shaft seals are found in rotating anode X-ray generators and in vacuum chambers used in the semiconductor industry. Ferrofluid seals are used in high-speed computer disk drives to eliminate harmful dust particles or other impurities that can cause the data-reading heads to crash into the disks. Another application of ferrofluids is in improving the performance of loudspeakers[3]. This is a material with which an experiment may be

[2][3] "Synchrotron Radiation." *Synchrotron Radiation*. N.p., n.d. Web. 30 Nov. 2016. "University of Wisconsin-Madison Materials Research Science and Engineering Cente r Interdisciplinary Education Group." *Ferrofluids*. N.p., n.d. Web. 14 Oct. 2016.

suggested. Not to be, however not mistaken with the liquid present in the core which is actual liquid metal.

The real question is more whether or not the field lines can change with this liquid potential. To answer this question we can look at two main things, whether the liquid is moving itself, or if there is an exterior force causing the liquid to move and whether the magnetic flux is penetrated. Addressing the concept of the geodynamo, there are convection currents -currents that form a repeated cycle[2]- occurring in the outer core. Plus, there is an exterior force, the coriolis force, generated by the earth that further causes the liquid to be in motion. Assuming both of these factors are true, the liquid potential should be more than enough to penetrate and break the lines of magnetic force. Figure 2 describes how this would look like on a smaller scale. It will be very difficult to make a prototype of the model since the conditions necessary for such a prototype to function is next to impossible. Despite this fact, it is still theoretically sound since the motion between the coil and the geodynamo is all that is really required for electricity to be produced.

The practicality of such a project may be a challenge. However, the benefits of such an undertaking outweighs the cost. Referring to the figure above, to make this work, we would need to build an intercontinental belt which would arguably be the biggest barrier of the entire process. It would have to stretch around along the equator and stay connected throughout. Considering the amount of money that the government of any country spends on a number of unfruitful research every year, the monetary expenditure for such an undertaking would not be too high, depending on the materials used to build it. Copper is the most preferable metal since it does not corrode, is non-magnetic, and easily joined but other conductive materials can also be used, like Aluminium for it's lightness or economic advantage, or graphite. The setting of the belt can be on air or underground and predictably, the general population would rather have it out of their immediate sight but keeping it underground may require more spending on the long run to keep the belt protected and would be harder to repair in case of any wear or damage. Setting the coil underground may also make the land more prone to other natural disasters like landslides. This environmental issues and threats will be taken care of if we use alternative forms of a coil, but if the belt is constructed, perhaps designing the coil in a different shape, like an oval, depending on natural threats, would be in our best interest.

The more loops a coil has, the more electricity it will produce. Similarly, the thicker a coil is, the more electricity the coil will produce. It will be challenging and expensive to make the coil have a number of loops since even making a single such loop would have heavy costs and possibly maintenance liabilities. Therefore instead of making a number of such structures, it would be much more efficient on our part to make it thicker since that also has a positive correlation with the amount of electricity produced as described by creative science in paragraph 9[3]. To take on such a project will take more than just being able to afford to make such a vast structure. It will require the input and cooperation from other countries. This however does not limit our scope

2 *Volcano World* . Oregon State University, n.d. Web. 30 Nov. 2016.

3 Electrical Generators. "MAKING AN ELECTRICAL GENERATOR." *The Creative Science Centre* . N.p., n.d. Web. 14 Oct. 2016.

to the possibility of a single idle structure around the globe, if this project is undertaken this way. Addressing the safety issues, the human population along with animals will face no significant threat of being overexposed to magnetic fields that can alter biological functions since no new fields are being introduced to the system as described by the US department of the interior[4]. The other issue that poses a challenge to wildlife would be the construction of the supposed intercontinental belt, however, alternatives to constructing something of that is discussed later.

To be more practical, the belt does not have to physically exist. According to Faraday's law, we do not require a circular belt across a field in order for this idea to come into action. As long as we do not have gaps between a conductor that covers the field completely, we will be producing currents. This means that the heavy lifting and the construction of an actual belt would not have to be undertaken with the hassle of maintenance. Instead, we could be using structures that already exist including sections of electric poles that are not used, with which we can pass through copper wires and use that in order to connect the wires and make a structure resembling a coil. This may seem questionable but evidently, we would not necessarily even have to make this coil earth-wide. According to the electromagnetic induction phenomenon, as long as the magnet or the coil is moving, electricity will be produced. Therefore, since the magnet is always in motion due to it's liquid state, the primary thing we need to be concerned with would be to make sure that we provide a region for the field lines to pass through. It may also be possible to use wifi routers in homes to regulate a constant surge of electrons, if they can be manipulated to have or create the field of a coil that would act like a copper coil surrounding the earth or a large region of the earth. This planet is in need of cleaner sources of energy, and it needs it fast. Therefore, I conclude that this alternative, although will require further research, would certainly be one viable option.

---

[4] "Does the Earth's Magnetic Field Affect Human Health?" US department of the Interior, 15 June 2016.

Selected Works

Bibliography:

"Faraday's Law."   *PhET*. University of Colorado Boulder, 26 Sept. 2016. Web. 14 Oct. 2016. <https://phet.colorado.edu/en/simulation/faradays-law>.

By Using a Flash Light as an Energy Source We Held the Flash Light at Different lengths While the NXT Recorded the Wave Lengths. "Sustainability, Energy, and Technology at Suffolk." *Sustainability*

Energy and Technology at Suffolk. Suffolk.edu, n.d. Web. 14 Oct. 2016.

<http://sites.suffolk.edu/akimmel/>.

"University of Wisconsin-Madison Materials Research Science and Engineering Center

Interdisciplinary Education Group." *Ferrofluids*. N.p., n.d. Web. 14 Oct. 2016.

<http://education.mrsec.wisc.edu/background/ferrofluid/>.

"Does the Earth's Magnetic Field Affect Human Health?" US department of the Interior, 15 June 2016. Web. <https://www2.usgs.gov/faq/node/2746>.

Electrical Generators. "MAKING AN ELECTRICAL GENERATOR." *The Creative Science Centre*. N.p., n.d. Web. 14 Oct. 2016. <http://www.creative-science.org.uk/gen1.html>.

"Electromagnetic Induction." *Electromagnetic Induction*. N.p., n.d. Web. 14 Oct. 2016.

<https://www.nde-ed.org/EducationResources/HighSchool/Electricity/electroinduction.htm>.

"CHAPTER 32." *ELECTROMAGNETIC INDUCTION*. N.p., n.d. Web. 14 Oct. 2016.

<http://teacher.nsrl.rochester.edu/phy122/Lecture_Notes/Chapter32/chapter32.html>.

"How Is Mains Electricity Produced?" N.p., n.d. Web.
<
http://www.bbc.co.uk/schools/gcsebitesize/science/add_ocr_pre_2011/electric_circuits/mains produ cedrev1.shtml>.

"Applications of Electromagnetic Induction." N.p., n.d. Web.

<http://physics.bu.edu/~duffy/PY106/Electricgenerators.html>.

"Making a Magnet." University of Leicester, n.d. Web. 01 Nov. 2016.

Selected Works

<https://www.le.ac.uk/se/centres/sci/selfstudy/mam9.htm>.

Volcano World. Oregon State University, n.d. Web. 30 Nov. 2016.

<
http://volcano.oregonstate.edu/oldroot/education/vwlessons/lessons/Earths_layers/Earths_layers7.html>.

"Copper Properties and Uses. Introduction." *Copper Properties and Uses. Introduction.* N.p., n.d.

Web. 30 Nov. 2016. <http://resources.schoolscience.co.uk/CDA/14-16/chemistry/copch0pg5.html>.

"Synchrotron Radiation." *Synchrotron Radiation*. N.p., n.d. Web. 30 Nov. 2016.

<https://universe-review.ca/R05-02-synchrotron.htm>.

Investigating the relationship between the sinusoidal patterns of a pendulum's position, velocity, acceleration, jerk, and jounce.

Selected Works

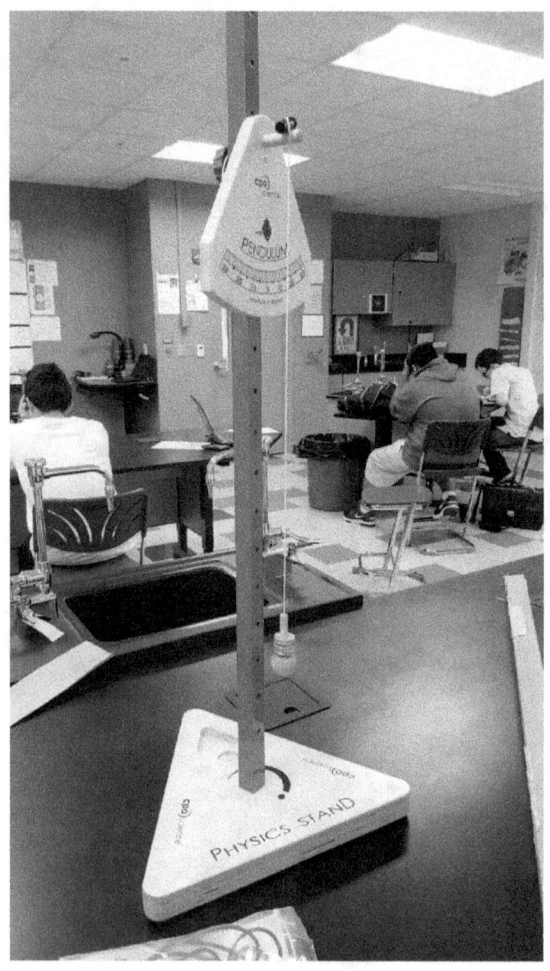

Zarrar Nashman

Background:

In physics, jerk, also known as jolt, surge, or lurch, is the rate of change of acceleration. The jerk of an object has numerous uses that deal with the comfort level passengers in a moving vehicle. The rate of change of Jerk is called jounce. All of these values are interconnected through derivatives. Suppose, the derivative of the position function of an object gives us the velocity function of that object. The second derivative will give us the function of the acceleration of that function and so on. As we keep on taking the derivative of the primary function, our fourth derivative gives us jounce. The reason I chose a pendulum to replicate motion is because plotting a pendulum's motion on the cartesian plane yields a trigonometric function like sine or cosine. And with sine and cosine functions, the fourth derivative theoretically leads back to the value of the original function. I always wanted to see if they graphically displayed the same pattern.

Aim: Investigating the sinusoidal patterns of jounce, jerk, acceleration, velocity, and displacement.

Hypothesis: The rate of change of jerk will form the same pattern as a displacement time graph.

Materials:

Stand

Clamp

Photogate

Selected Works

String

Pendulum bob

Wooden compass

Variables:

Manipulated variable-length of string

Responding variable-period

Control-same string used

-same protractor used

-mass of bob

-angle of release

Procedure:

Set up the clamp to the stand and attach the photogate below the wooden protractor. Attach the string to the ball and attach the top of the string to the top of the wooden extension and adjust the string for getting a string length of 35cm, making sure the bob passes in between the probes of the photogate. Hold the string from a 20 degree angle and then release the pendulum bob and measure the time it takes to make one complete oscillation using the photogate. Repeat the same steps with lengths 40cm, 45cm, 50cm, 55cm, 60cm. Then calculate for jerk, acceleration, and velocity using formulas.

Diagram:

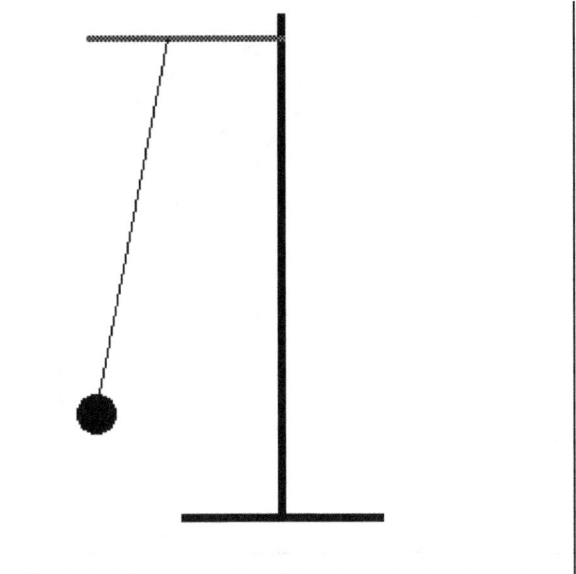

Data:

length	T1	T2	T3	T4	T5	T6	T7	T8	T9	T10	mean
35	4.04	4.03	4.04	4.04	4.04	4.04	4.04	4.05	4.04	4.04	4.04
40	4.37	4.37	4.37	4.36	4.37	4.37	4.38	4.37	4.38	4.37	4.37
45	4.60	4.60	4.60	4.60	4.60	4.59	4.60	4.61	4.60	4.60	4.60
50	4.87	4.87	4.86	4.87	4.87	4.86	4.87	4.87	4.87	4.87	4.87
55	5.11	5.12	5.12	5.12	5.11	5.12	5.12	5.12	5.11	5.12	5.12
60	5.37	5.36	5.35	5.36	5.35	5.36	5.35	5.36	5.35	5.37	5.36

Analysis:

Graphically representing the motion of a pendulum will always produce a sinusoidal curve. Using the formula below, we can calculate the arclength that gives us the total distance the pendulum bob has travelled at these different lengths:

$$Arc\,length = \frac{Angle}{360°} \times \pi d$$

Where d stands for the diameter. Using this formula, we get:

length(cm)	Arc length(cm)
35	6.11
40	6.98
45	7.85
50	8.73
55	9.60
60	10.47

If we were to graphically represent the displacement pattern of the graph, we would get:

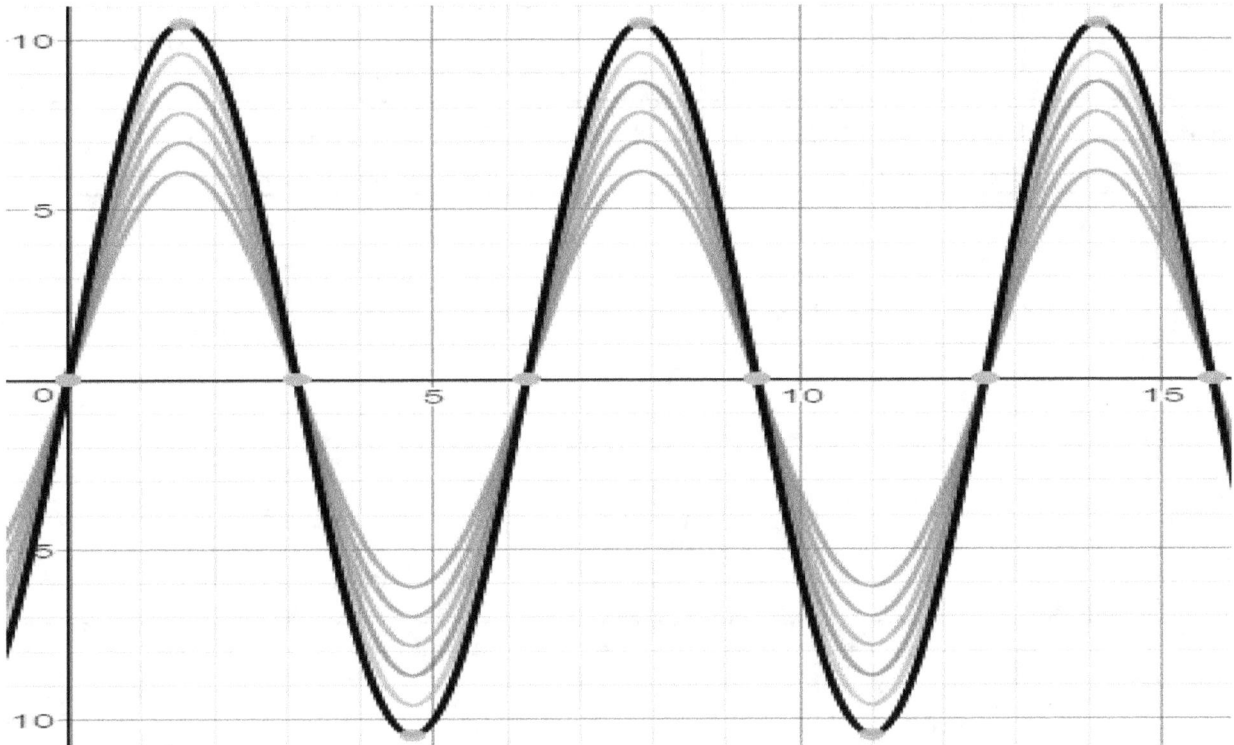

https://www.desmos.com/calculator

The graphical plotting of the motion will produce a sin function as shown above. This graph displays our pendulum's displacement with respect to time. In order to display the rate of change of jerk time curve, we will apply the formula:

$$\frac{y_2 - y_1}{x_2 - x_1}$$

This is the general formula for the rate of change. Since we know that the jerk is the rate of change of acceleration, which is the rate of change of velocity, which is the rate of change of displacement, we have to keep on applying this formula until we get to the rate of change of jerk. The next step would be to apply this formula to our primary numbers, which is displacement in order to get our velocity time graph. Applying the formula, we get:

position	6.11	6.98	7.85	8.73	9.60	10.47
velocity	1.51	1.60	1.71	1.79	1.88	1.96

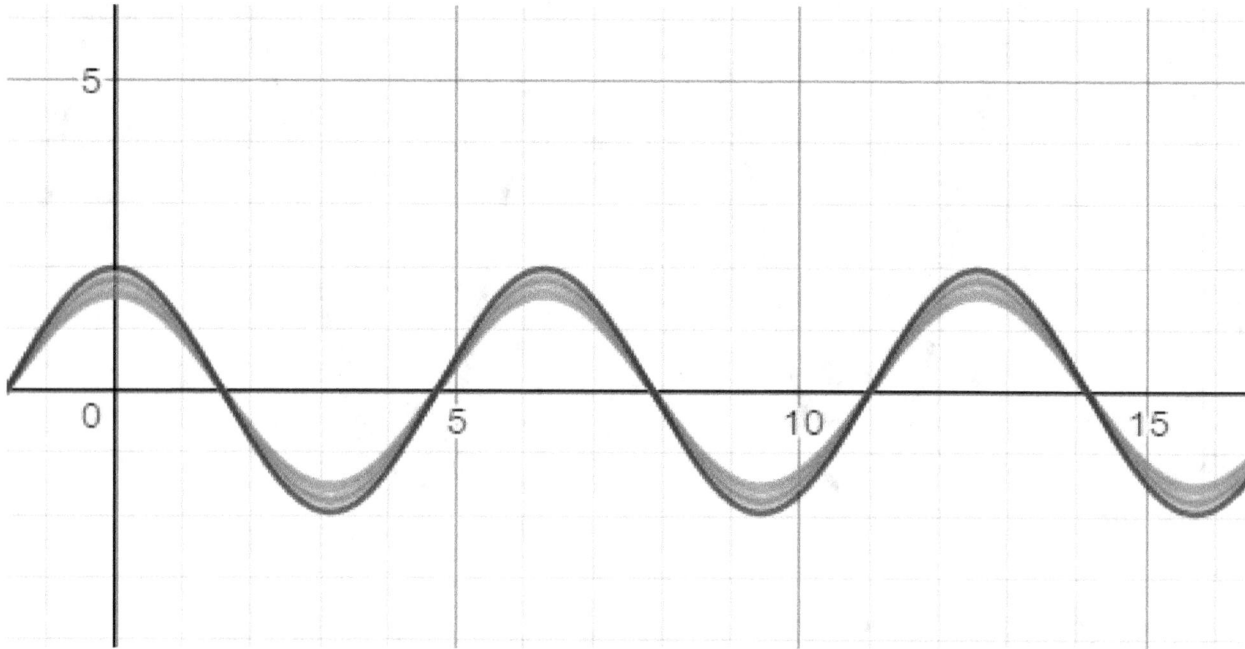

Note that the amplitude is decreasing for every step we take. However, the pattern still remains similar with a horizontal shift and a vertical dilation. Moving on to the acceleration time graph:

velocity	1.51	1.60	1.71	1.79	1.88	1.96
acceleration	0.38	0.36	0.37	0.37	0.37	0.37

Selected Works

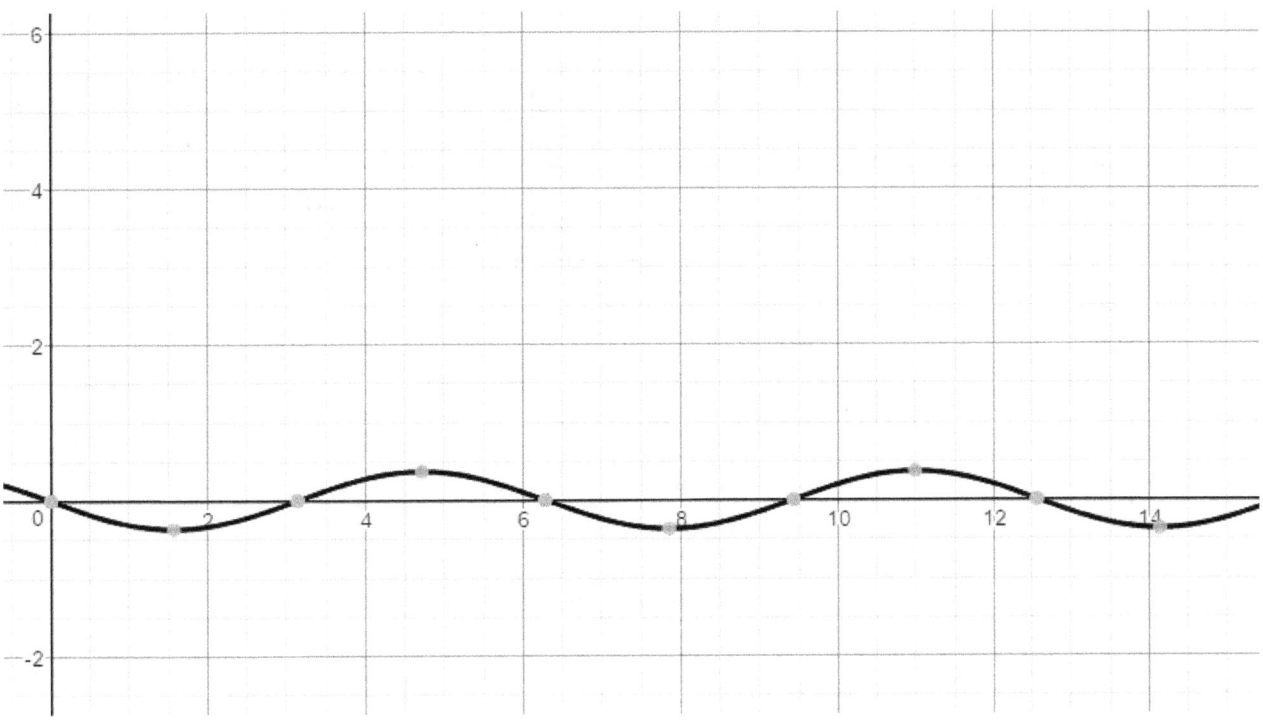

acceleration	0.38	0.36	0.37	0.37	0.37	0.37
jerk	0.09	0.08	0.08	0.08	0.07	0.07

jerk	0.09	0.08	0.08	0.08	0.07	0.07
jounce	0.02	0.02	0.02	0.02	0.01	0.01

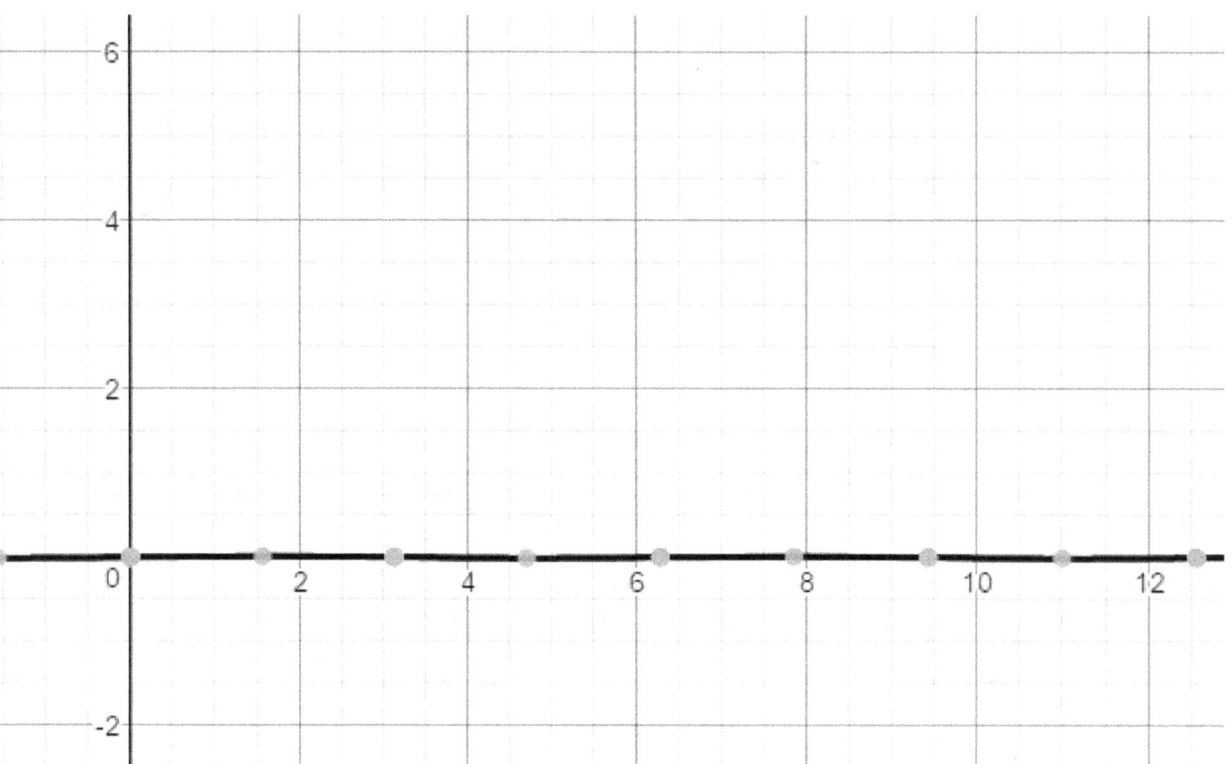

Conclusion:

In the acceleration time curve and forward, only the final line seems to be visible. This is because the values are merging closer and closer to each other as we go higher up. Zooming in, however, we could magnify the difference by using a more precise instrument and taking into account a couple of more decimal places. Comparing the displacement time curve and the jounce time curve clearly shows that the pattern created by both are almost completely identical. The only difference being is that the curves have different amplitudes. The periods of the curves are not correct as it was following standard trigonometric values which should have been replaced by the measured time. Moreover, even though the instrument the data was collected with was more precise, I only took into account the raw data upto two decimal places, which later caused a problem with significant figures when calculating the data points for the jounce time graph. This also would not allow me to zoom in and show the slight changes in graphs since most of the values rounded up to the same value. Furthermore, generating the graphs through an online source did not allow me to insert error bars.

Bibliography:

"Arcs and Sectors." BBC, n.d. Web.
<http://www.bbc.co.uk/education/guides/ztg9q6f/revision>.

Roberts, Donna. "Slope and Rate of Change." *Regentsprep.com*. N.p., n.d. Web. 03 Nov. 2016.
<http://www.regentsprep.org/regents/math/algebra/ac1/rate.htm>.

"What Is the Term Used for the Third Derivative of Position?" *Third Derivative of Position*. Ed. Stephanie Gragert. N.p., n.d. Web. 03 Nov. 2016.
<http://math.ucr.edu/home/baez/physics/General/jerk.html>.

"What Is Derivatives Of Displacement?" *What Is Derivatives Of Displacement?* Wearcam.org, n.d. Web. 03 Nov. 2016. <http://wearcam.org/absement/Derivatives_of_displacement.htm>.

Simanek, Donald E. "Simple Pendulum." *lhup.edu*. N.p., 2004. Web. 03 Nov. 2016.
<https://www.lhup.edu/~dsimanek/scenario/labman1/pendulum.htm>.

Investigating the effect of magnesium on photosynthesis

Aim:

Investigating the impact of magnesium in the process of photosynthesis.

Hypothesis:

The concentration of magnesium is directly proportional to the rate of photosynthesis.

Introduction:

When I was in middle school, my teacher taught us how different ions affected different parts of the plant and one of those prime ions was magnesium. I had always imagined since magnesium was responsible for making new chlorophyll, as to how the functions associated with magnesium would affect the photosynthesis rate of a plant.

Background Information:

http://www.easy-grow.co.uk/using-magnesium-to-boost-photosynthesis/

Magnesium is the central chemical in the making of chlorophyll, the green pigment in plants. Magnesium is so important to photosynthesis that if there is a magnesium deficiency, plants will remove magnesium ions from the chlorophyll in the lower leaves, and translocate it to the upper leaves where the plant needs it the most. That is why magnesium deficiency appear as interveinal chlorosis in the lower leaves before leaves from any other sections. The veins remain green, but the leaf tissue turns yellow from the lack of magnesium, which happens to be an indicator of magnesium deficiency.

https://www.hydroponics.net/learn/deficiency_by_element.php

Magnesium poisoning is extremely rare and does not exhibit any physical appearance.

The reason jalapeno plants were chosen for this experiment is because they are easy to grow and most of their nutritional deficiencies are easy to locate as almost all of them cause yellowing of the leaves.

Selected Works

Materials:

-Beaker(20ml)

-Bell jar or any sealed, transparent, and relatively airtight container.
-Jalapeno plants(10 weeks of maturity)X6

-Potting soil
-6 plastic flowering pots
-Tap water
-Magnesium Sulfate fertilizer

-Vernier Instrument Oxygen meter

-Marker
-Rubber tube

Variables:

Responding variable: oxygen concentration
Manipulated variable: amount of fertilizer applied

Control: Plant without fertilizers added

Same amount of water administered

Same length of exposure to light

Exposure to the same temperature at all times.

Procedures:

Label each of the pots with the desired amount of fertilizer being tested(0mg, 5mg, 10mg, 15mg, 20mg, 25mg) using a marker on all of the plants Obtain six jalapeno plants with a 10 week maturity and place each of the plants in each of the pots. Add potting soil until the brim of the plastic flower pot is filled. All the leaves should be visible above the soil. Add the desired amounts of fertilizer labelled on each of the pots to each of the plants. Water the plants regularly for a period of seven days. After seven days have passed, obtain a bell jar or any other airtight material to make sure that there are no gaseous exchange with the surrounding. For the following experiment, an airtight shoebox was used. To make an airtight shoebox, first cut out a hole from a standard cardboard shoebox, slightly smaller than the oxygen sensor's mouth. Then cut out a square from the resealable side of the box and replace it with transparent plastic film firmly attached to the box as described in figure 1. Seal all the openings with duct tape except for the main swinging side. Make sure that the oxygen meter is attached to the vernier device and that the vernier device is plugged into a power source. Place the sensor into the circular opening created and place one plant inside of the shoebox through the swinging side in the presence of light and close it. Start measuring the concentration of oxygen at the beginning of the trial and record the oxygen concentration after 80 seconds. Subtract the initial amount from the final amount and it will produce the amount of oxygen produced in 80 seconds. Repeat the same step for all the plants with 9 trials each.

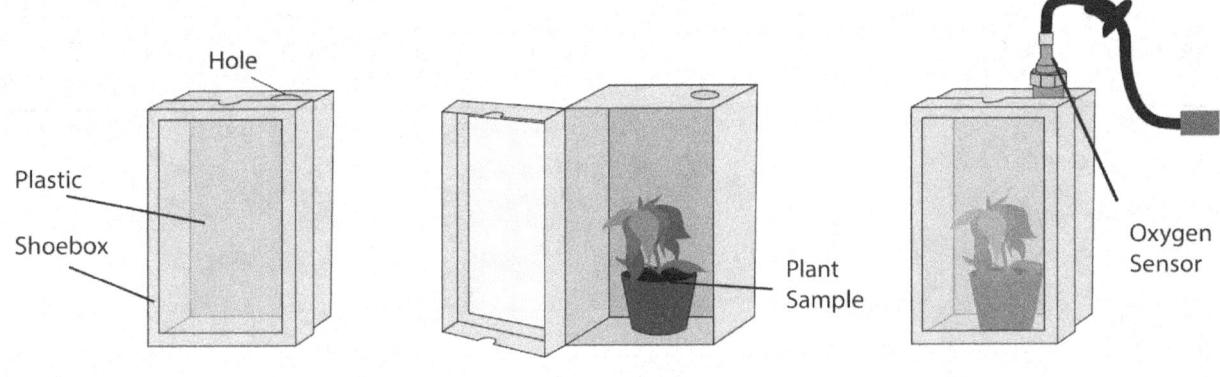

Observation:

The Plants with the two highest doses appeared to be dying. The leaves of those plants also appeared to be wilting.

Data:

	Trial 1	Trial 2	Trial 3	Trial 4	Trial 5	Trial 6	Trial 7	Trial 8	Trial 9	Mean

0mg	0.00	0.03	0.05	0.02	0.01	0.00	0.02	0.03	0.02	0.02
5mg	0.07	0.08	0.05	0.09	0.02	0.08	0.09	0.07	0.02	0.063
10mg	0.05	0.06	0.05	0.03	0.04	0.06	0.02	0.01	0.01	0.036
15mg	0.03	0.02	0.01	0.01	0.02	0.00	0.01	0.00	0.02	0.013
20mg	0.00	0.00	0.02	0.00	0.00	0.01	0.00	0.01	0.00	0.004
25mg	0.00	0.00	0.00	0.00	0.00	0.00	0.00	0.00	0.00	0.0

Analysation:

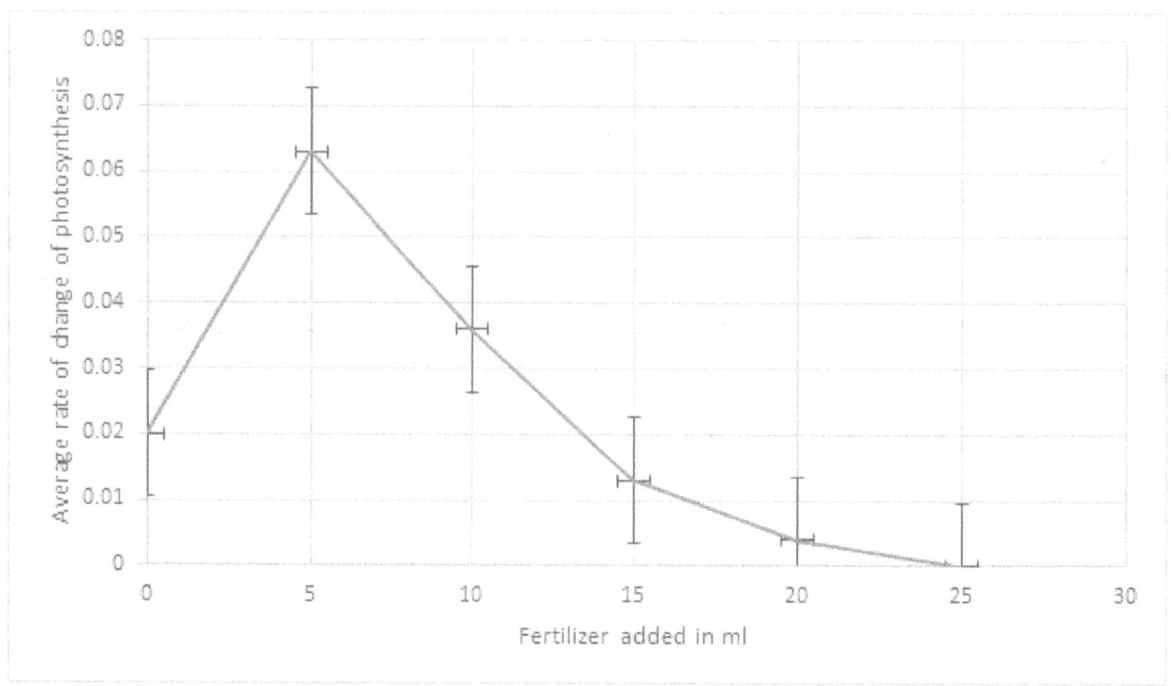

For the samples with 0ml and 5 ml fertilizers, a steady growth was observed and the rate of oxygen production was almost zero for the last two sample. They appeared to be dead and wilting.

Conclusion:

It is evident from the the experiment that at low doses, the magnesium sulphate fertilizer does have a positive correlation with the rate of photosynthesis. However, at the two highest doses, we have seen that the photosynthesis rates plummet. This was surprising as magnesium toxicity, as mentioned earlier, is very rare and does not exhibit any other visible physical symptoms.

Error analysis:

This experiment had a number of possible errors. The box used instead of a bell jar was not completely sealed as there was space between the flaps still available. A bell jar is highly recommended for this experiment. Due to time constrictions, I was forced to record the plants' data at night using an artificial light source at night. That most likely impacted the readings on the oxygen meter since the light was provided at an odd time with respect to the plants' biological cycle since they respire at night. Repeating the experiment during the daytime may have larger changes and better readings.

Bibliography:

By Easy Grow Ltd on Jun 24, 2015 | 0 Comments. "Magnesium for Photosynthesis | Easy Grow Ltd." *Easy Grow Ltd*. N.p., 04 July 2015. Web. 14 Oct. 2016.

"Symptoms of Deficiencies and Toxicities - Greentrees Hydroponics."*Symptoms of Deficiencies and Toxicities - Greentrees Hydroponics*. Hydroponics.net, n.d. Web. 14 Oct. 2016.

"Magnesium Deficiency." *Magnesium Deficiency*. Lucidcentral.org, n.d. Web. 14 Oct. 2016.

I think my actual journey to where I am now started when I was in middle school. I was never a very good student before I was fourteen years old, and my grades were mostly average. What would a person possibly want to do other than play video games and hang out with friends, was my thought at the time. I was deluded into my own view of life through the nurturing of my parents. We were not well off by any means, but they had enough to make sure that I did not have much youthful envy inside of me. I was, however, always lectured about how important education is and they probably never stuck to me for more than a week. My teachers, on every parent-teacher meeting day used to tell my parents that I was an extremely bright kid, but I need to work more. I knew that I would perhaps easily be one of the best students. After sixth grade ended, my grades reached a record low. This caused many of my relatives to lose faith in me, faith that I would someday become something great. What many people did not know was that I was the only child of my parents and the only heir to my family, since my uncles or aunts never married or never had kids. So it was my responsibility to make my family's name shine. I was already stressed with my parent's constant reminder of such expectations. On top of that, my grades caused me to lose all hope and interest in anything to do with education. Even the closest people to me lost faith in me. I can't say I never had thoughts about how it would be so much easier and less painful for others if I never existed in the first place. No regrets, and no disappointments. But after the summer ended that year, I picked up my much needed spiritual motivation. I had to believe that even if the world did not believe in me, God's belief would be sufficient for me. The only person that believed in me in the heavens was The Almighty. No one else believed in me on earth. So I believed in myself. As seventh grade started, I made a commitment to myself that I would not hang out with my friends or play any video games this year. It was going very well, and after my midterms, I was slowly lurching atop the leaderboards of my class. Along with it, I developed a habit of studying for at least two hours every day. When my finals appeared, I knew that I could not afford to just settle on average, I had to do well. Not for anyone else, not anymore, but for myself. And I kept myself in complete isolation inside of our home library for a period of two days, trying to focus on the goal, when suddenly something happened. I had a realization that everything that I lived on were my parent's. If I don't actually do well, my value in this world would be insignificant. I never wanted to end up like that, never in my deepest imaginations have I thought of it that way. Right after, my mind shifted to a focus that I have never had before. I was going through all my textbooks in a day. I was sure I felt that I had the spirit of God. By the time finals came, I was ready. I was ready to knock it out of the park but nervous too. I never actually sat down for an exam with that much knowledge. By the end of the year, I placed 13th in my whole batch of 200 students. The next year was similar and I was thoroughly taking notes in class and doing all my homework everyday. By the end of the year, I received a certificate of honors from my school principal for my exemplary grades. I did not even care about what people thought about me anymore, all I knew was that I needed to climb this ladder. My freshman year started soon afterwards and that was when I realized that I don't study through obligation anymore, but choose to do it through my free will. I was way past the majority of my classmates when it came to academic standings. It was also the first year we had electives, and I chose the toughest courses possible including higher mathematics. It was a tough year, but I powered through it and ended up with the second best grades in class. Sophomore year was when I decided to take my challenge to a different level. I was taking two AP classes and one pre ib class and I passed all of the exams. After that year, I realized that education was the ultimate goal. Regardless of what we have in our lives, they are bound to leave us some time. Knowledge was the only thing that would provide for you when in time of need. I also picked up a keen interest in neurology several years before that, but decided that I would become a neurosurgeon at the end of my journey. Doing this for myself was great but I owed the world it's fair share of my merit. The world had to know what I had to offer. Even a single life saved or prolonged with my knowledge would be enough to suffice for all the hard work I had done thus far and will continue to do in the future. With that in mind, I decided to go

Selected Works

into the IB diploma program which spans over a period of two years. The two years were filled with more hard work and it is almost time for me to sit for the final tests. After all this, I stand as one of the best students in school, Varsity soccer player, and voted by the school as the most likely person to succeed in life. My learning will not stop there, actually it is just the beginning. I will reach my goal because I am the revolution the world is waiting for. I will be the next big thing in neuroscience and in time, I will help others.

And whoever saves one[life] - it is as if he had saved mankind entirely.
Quran 5:32

www.ingramcontent.com/pod-product-compliance
Lightning Source LLC
Chambersburg PA
CBHW080859170526
45158CB00009B/2775